NATURE'S GROSSEST

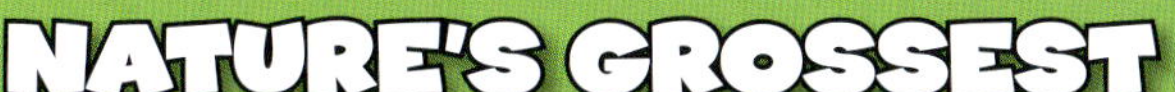

HIPPOS THROW POOP!

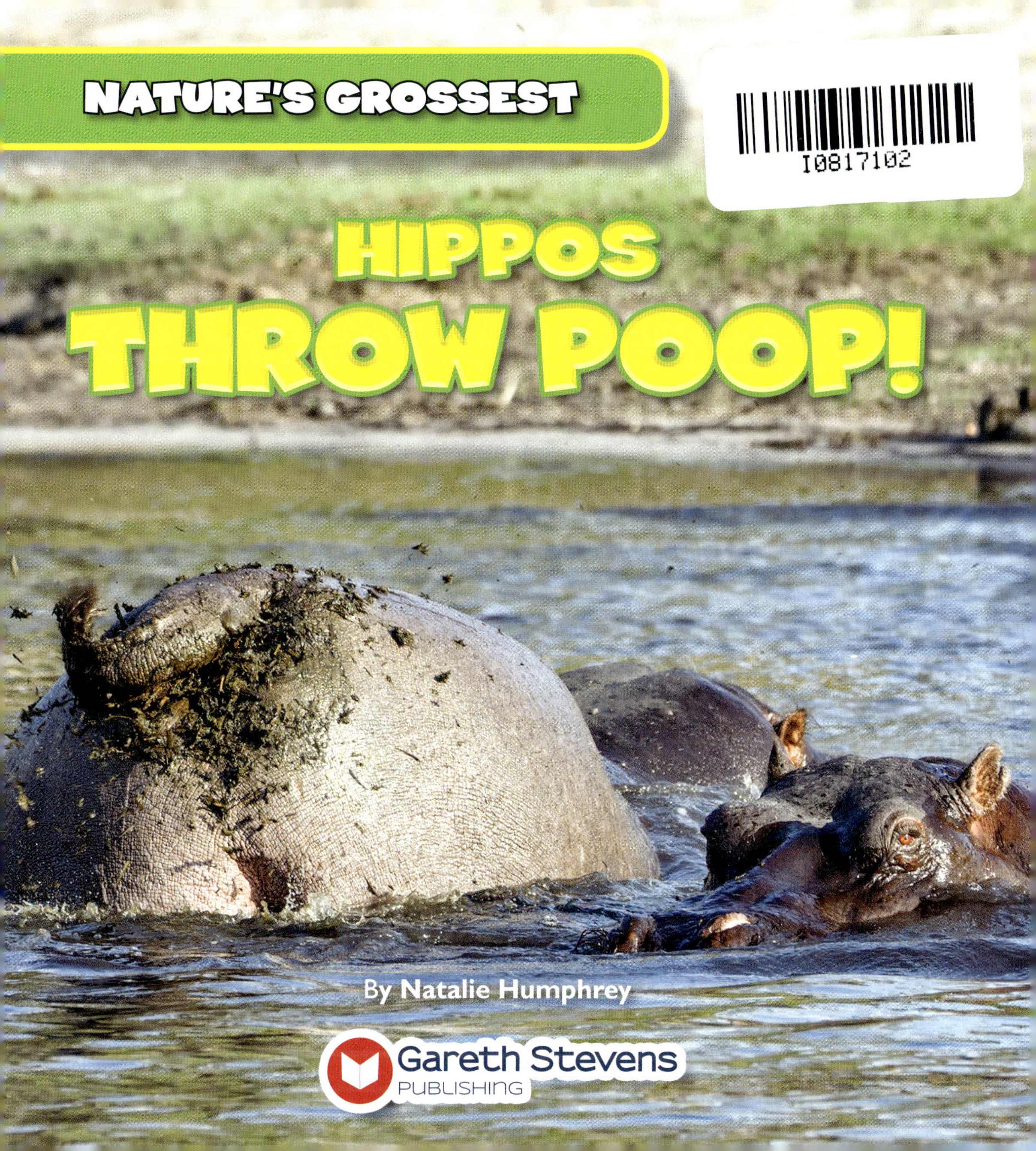

By Natalie Humphrey

Gareth Stevens PUBLISHING

Please visit our website, www.garethstevens.com. For a free color catalog of all our high-quality books, call toll free 1-800-542-2595 or fax 1-877-542-2596.

Library of Congress Cataloging-in-Publication Data
Names: Humphrey, Natalie, author.
Title: Hippos throw poop! / Natalie Humphrey.
Description: Buffalo, New York : Gareth Stevens Publishing, [2024] | Series: Nature's grossest | Includes index.
Identifiers: LCCN 2022054147 (print) | LCCN 2022054148 (ebook) | ISBN 9781538285725 (library binding) | ISBN 9781538285718 (paperback) | ISBN 9781538285732 (ebook)
Subjects: LCSH: Hippopotamus–Juvenile literature. | Animal droppings–Juvenile literature.
Classification: LCC QL737.U57 H86 2024 (print) | LCC QL737.U57 (ebook) | DDC 599.63/5–dc23/eng/20221115
LC record available at https://lccn.loc.gov/2022054147
LC ebook record available at https://lccn.loc.gov/2022054148

Published in 2024 by
Gareth Stevens Publishing
2544 Clinton Street
Buffalo, NY 14224

Designer: Leslie Taylor
Editor: Natalie Humphrey

Photo credits: Series art (background) Oleksii Natykach/Shutterstock.com; cover, p. 19 Claude Huot/Shutterstock.com; p. 5 nataliatamkovich/Shutterstock.com; p. 7 BlueOrange Studio/Shutterstock.com; p. 9 Yakov Oskanov/Shutterstock.com; p. 11 cloudzilla/Flickr.com; p. 13 Mogens Trolle/Shutterstock.com; p. 15 PhotocechCZ/Shutterstock.com; p. 17 Jukka Jantunen/Shutterstock.com; p. 21 Jens Goos/Shutterstock.com.

Printed in the United States of America

CPSIA compliance information: Batch #CSGS24: For further information contact Gareth Stevens at 1-800-542-2595.

CONTENTS

Boldface words appear in the glossary.

Flinging Poop!

Usually weighing from 2 to 4 tons (1.8 to 3.6 mt), a hippopotamus is a giant **mammal**! Hippos have a strange way of telling other animals to back off: they fling, or throw, their poop!

At Home in Africa

Hippos are found in **sub-Saharan** Africa. They spend most of their time in the water and only come out for food. When they need to, hippos will travel 6 miles (9.6 km) to find their dinner!

Made for Water

A hippo's body is perfect for life in the water. Hippos spend most of the day in their lake and river homes. A hippo's eyes and nose are on top of its head, which helps it breathe and stay mostly **submerged**.

Life Underwater

Even though hippos spend most of their time in the water, they can't swim. Hippos walk underwater! Hippos will also push off objects underwater and **glide** for a short time. A hippo can hold its breath for around five minutes!

Awake at Night

Since it's so hot during the day, hippos usually eat at night. Hippos eat around 88 pounds (40 kg) of food per night. Hippos are **herbivores.** They eat grasses and fruit around the water they live in.

Ready to Fight

Hippos only eat plants, but they have huge teeth! These big teeth are mostly used to scare off **threats** and fight. Hippos can be aggressive, or ready and willing to fight! In Africa, hippos cause around 500 human deaths per year.

Poop Marking

Hippos usually only attack, or fight, when another animal comes into their water **territory**. To mark their territory, hippos use their poop! While a hippo is pooping, it will swing its tail around to fling the poop in every direction.

Stranger Danger!

Hippos fling their poop if a strange hippo gets too close. This is to let the other hippo know not to get any closer or there might be a fight. If one hippo hears another hippo, they'll fling their poop!

Watch Out!

Hippos can fling their poop around 33 feet (10 m), but sometimes their poop goes even farther. Some people visiting Africa have even been hit by a spray of hippo poop. Now, that's gross!

GLOSSARY

glide: To move in a smooth and graceful way.

herbivore: An animal that eats only plants.

mammal: A warm-blooded animal that has a backbone and hair, breathes air, and feeds milk to its young.

submerge: To put or sink below the surface of water.

sub-Saharan: The region in Africa below the Sahara, a large desert.

territory: An area of land that an animal considers to be its own and will fight to defend.

threat: Something likely to cause harm.

FOR MORE INFORMATION

BOOKS

Brandle, Marie. *Hippopotamus Calves in the Wild.* Minneapolis, MN: Jump!, Inc. 2023.

Riggs, Kate. *Hippopotamuses.* Mankato, MN: Creative Education/Creative Paperbacks 2022.

WEBSITES

National Geographic Kids
kids.nationalgeographic.com/animals/mammals/facts/hippopotamus
Check out maps, videos, and more photographs of hippos in the wild.

San Diego Zoo Wildlife Explorers
sdzwildlifeexplorers.org/animals/river-hippo?keys=hippo
Learn more fun facts about hippos and how they live.

Publisher's note to educators and parents: Our editors have carefully reviewed these websites to ensure that they are suitable for students. Many websites change frequently, however, and we cannot guarantee that a site's future contents will continue to meet our high standards of quality and educational value. Be advised that students should be closely supervised whenever they access the Internet.

INDEX